"Once you have tasted flight, you will forever walk the world with your eyes turned skyward. For there you have been and there you will always long to return."

Leonardo Da Vinci

BUILD IT YOURSELF!
PVC ROCKET ENGINE

A do-it-yourself guide for building a K450 PVC plastic rocket engine that will propel a rocket over 5000 feet.

Dan Pollino

This book is dedicated to my wife Lisa and my good friend Alberto Gassol. Without their support through countless tests, technical discussions and launches, the K450 PVC rocket engine would still be a shoddy pencil drawing in my sketchbook.

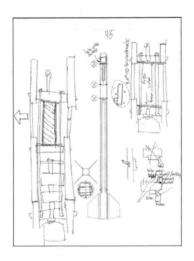

Contents

Disclaimer

By purchasing and using the PVC Rocket Engine book, I hereby fully and forever release and discharge indemnify and hold harmless Inverse Engineering and its employees and agents from any and all liabilities, claims, demands, damages, rights of action, suits or causes of action present of future, whether they same be known or unknown, anticipated or unanticipated, resulting from or arising out of my use or intended use of said PVC Rocket Engine book. I fully and forever release and discharge Inverse Engineering and its employees and agents from any and all negligent acts and omissions in the same, and intend to be legally bound by this release. By purchasing and using the PVC Rocket Engine book I therefore assume all risk associated with using the PVC Rocket Engine, even if they arise from the negligence of Inverse Engineering. My participation in this activity is voluntary and no one is forcing me to participate in spite of the risks. I understand the effect of this waiver and acceptance of risk on my legal rights.

This book, like all scientific books, is based on information acquired from many sources. Books, websites and tedious research have all contributed. In addition to these sources, personal experience and communication with other rocket enthusiasts all over the world have greatly added to the information presented.

Safety is and should always be the primary concern when mixing and using low explosives. Highly energetic compounds pose the risk of serious injury if precautions are not taken and common sense safety measures not adhered to consistently.

Icon representations of common OSHA Caution and Danger signs have been added to each step in this manual to make the reader immediately aware of any risk.

Since 9-11, many additional government restrictions have been added to the manufacture and use of low explosives. Especially those containing potassium nitrate composite mixtures as detailed in this manual. A link to Federal explosives law has been added in the *list of links* to assist the reader with the proper forms and procedures needed to make and use the K450 PVC rocket engine legally. This information is correct to the best of the author's knowledge at the time of this writing. Since these rules and regulations are constantly changing, the author urges the reader to research this topic thoroughly on their own. Check with the authorities in your area regarding local testing and launching restrictions.

As the author of this manual, I reserve all rights to the textual, graphical and other content appearing in this manual. Redistribution or commercial use is prohibited without express written permission by the author. Information in this manual is provided "as is" without warranty of any kind, either express or implied. In no event will the author be liable for damages of any kind. While every effort is made to assure the correctness of information, this manual may inadvertently contain technical inaccuracies or typographical errors. Information may be changed or updated without notice. The author may also make improvements and/or changes to the products and the programs described in this manual at any time without notice. The author assumes no responsibility for errors or omissions in this publication or other documents which are referenced by or linked to in this publication.

The purpose of this manual is to illustrate the fabrication of a very specific type of rocket engine, the 2" PVC K450 engine. Even though the instructions contained in this manual describe structural fastening points that are used to connect the engine to a rocket body, general rocket engine dynamics, rocket construction and rocket flight dynamics are not within the scope of this manual. It is the reader's responsibility to research these areas elsewhere.

About McMaster-Carr

McMaster-Carr (www.mcmaster.com) is an industrial supplier based in Los Angeles, Atlanta, Chicago, Cleveland and New Jersey. Many items in this manual are referenced as McMaster part numbers to ensure the same parts and tools that are used by the author are available to the reader. **Many tools used by the author can be substituted for less expensive versions or different tools altogether. The reader is encouraged to experiment with their own methods to decrease costs and improve productivity.**

About Potassium Nitrate, Kno3

Potassium Nitrate, Kno3, is the oxidizing component in the K450 PVC fuel mixture. It is a white powder that has a wide variety of commercial and industrial applications. It can be purchased at any local chemical supplier or online, see *List of Links*.

About Powdered Sugar

Powdered sugar, also known as confectioner's sugar or icing sugar is used in food production where a quick dissolving sugar is required. It is the main fuel component in the K450 PVC fuel mixture and can be found at most food stores.

About Corn Syrup

Corn syrup is made from corn starch and composed mainly of glucose. It is used to soften texture, add volume, prohibit crystallization and enhance flavor in foods. Corn Syrup is used in the K450 PVC engine as a minor fuel and an element that adds flexibility to the fuel. It can be found at most food stores.

About K450

The 'K' in K450 is a measurement of the total impulse of the engine and the '450' is a measurement of the average thrust.

Average thrust is a measure of how slowly or quickly the motor delivers its total energy, and is measured in Newtons. Total impulse is a measurement of the overall energy contained in a rocket motor and is measured in Newton seconds. The letter "K" designation means that there is anywhere from 1280 to 2560 Newton seconds of total impulse available in this motor.

The K450 has an average thrust of 473 Newtons and a total impulse of 2104 Newton seconds.

K450 PVC Rocket Engine Fuel Specifications

Fuel Composition
The fuel for the K450 PVC rocket engine is comprised of Potassium Nitrate (Kno3), powdered sugar and corn syrup in the following proportions:

Kno3   65%
Powdered sugar   16%
Corn Syrup   19%

Fuel Weight
There is 3.95 lbs. of fuel in each engine.

Burn Time
Each engine burns for 2.75 seconds once ignited.

Thrust
Initial thrust of the K450 engine is 100 pounds with a maximum thrust of 300 pounds.

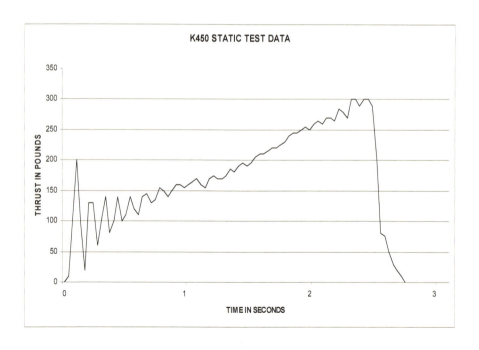

List of Links

Inverse Engineering
www.inverseengineering.com

My website that catalogs nearly ten years of rocketry experiments.

Propulsion Experimental
www.intertlan.com/cohetes

The website of my good friend Alberto Gassol.

Richard Nakka's Experimental Rocketry
www.nakka-rocketry.net

The website of Richard Nakka, a pioneer in sugar based propellants.

McMaster Carr
www.mcmaster.com

Los Angeles based industrial supplier. Many tools and materials referenced in this manual are available here.

PVC Only
www.pvconly.com

Online supplier of Kno3 (Potassium Nitrate) as well as other rocket supplies.

Skylighter
www.skylighter.com

Online supplier of Kno3 (Potassium Nitrate) and many other pyrotechnic chemicals not used in this manual.

Sugar Propellant List
http://rocketeers.com/mailman/listinfo/sugpro

An internet mailing list for people interested in sugar based propellants.

ATF
www.atf.gov

Bureau of Alcohol, Tobacco, Firearms and Explosives. The federal regulating body for low explosives such as the fuel used in the K450 PVC engine.

ATF Orange Book
http://www.info-central.org/regulatory/orangebook/

Federal Explosives Law and detailed instructions on how to obtain your Low Explosives Users Permit (LEUP)

FAA
www.faa.gov

Forms and regulations from the Federal Aviation Administration.

Adobe Acrobat Reader
www.adobe.com/products/acrobat/
readstep2.html

Adobe Acrobat Reader used to open and print out templates needed for the construction of the K450 nozzle.

Wikipedia Tap & Die Tutorial
www.wikipedia.org/wiki/Tap_and_die

Tapping tutorial.

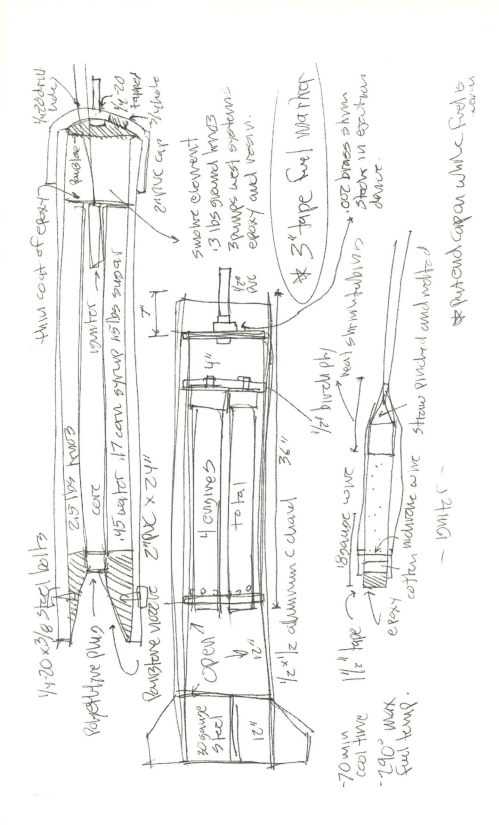

Chapter 01

MAKING THE NOZZLE

Besides the fuel, the nozzle is the most important part of the K450 PVC Engine. A heavy steel washer ensures that the throat will not erode and a 15 degree divergent cone helps increase the speed of the exhaust gasses for maximum performance. Even though the tolerances are very tight, it's very easy to make.

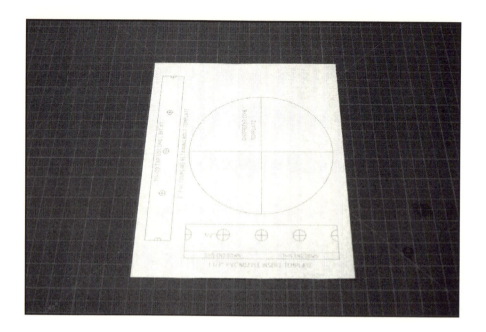

Step 01

Download and print out a copy of the *K450 Nozzle Template* from www.inverseengineering.com. All the templates for this project can be found under the download templates tab on the main page. Any kind of paper will work but heavier card stock type papers will hold up better. Since PVC pipe and fittings vary in size slightly by manufacturer and greatly internationally, these drawings may need to be altered to fit the outside diameter of the pipe and fittings that are locally available to you.

Materials
- 8.5" x 11" Paper

Tools
- Adobe Acrobat (see *List of Links*)
- Printer

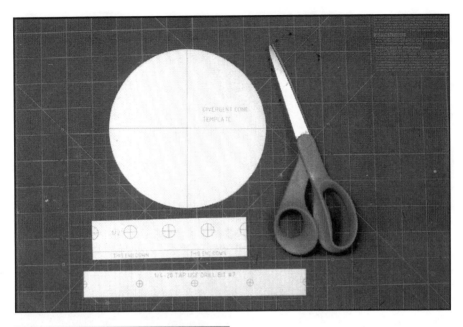

Step 02
The *K450 Nozzle Template* contains three templates.
- Divergent cone template.
- 2" PVC Coupling retaining bolt template.
- 1 1/2" PVC Nozzle insert template.

Cut out all three templates as accurately as possible.

Materials
- K450 Nozzle Template

Tools
- Scissors

Step 03
Cut the *Divergent cone template* along the red line. Cut along one radius of the circle, from the edge of the circle to the center of the circle.

Materials
- K450 Nozzle Template

Tools
- Scissors

Step 04
Cut a piece of 1 1/2" Schedule 40 PVC pipe 1/2" long.

Materials
- 1 1/2" Schedule 40 PVC pipe

Tools
- Electric miter saw or PVC pipe cutter

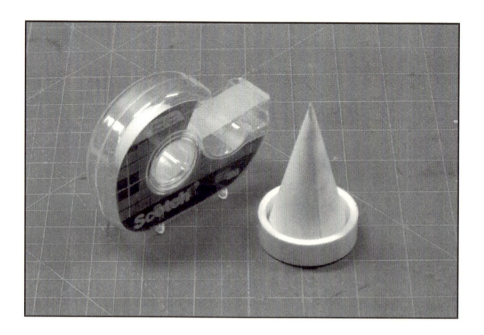

Step 05
Roll the *Divergent cone template* into a cone until the large end fits inside the 1 1/2" x 1/2" PVC ring. Let the cone open inside the PVC ring until it fits snuggly inside the ring. Use a piece of scotch tape on the cone seam to prevent the cone from increasing in size when the PVC ring is removed. Remove the PVC ring

Materials
- Scotch tape

Tools
- None

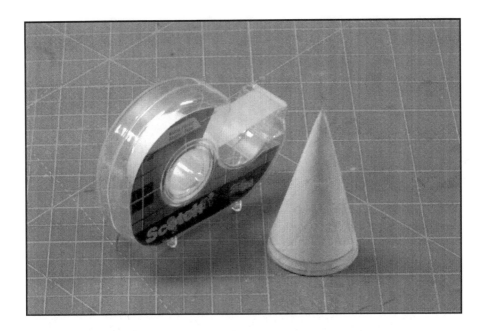

Step 06
. Using several pieces of scotch tape, cover the entire exterior of the cone with a layer of scotch tape. This will prevent the hydraulic cement from sticking to the paper and will aid in the cone's removal from the nozzle. Set the cone aside.

Materials
- Scotch tape

Tools
- None

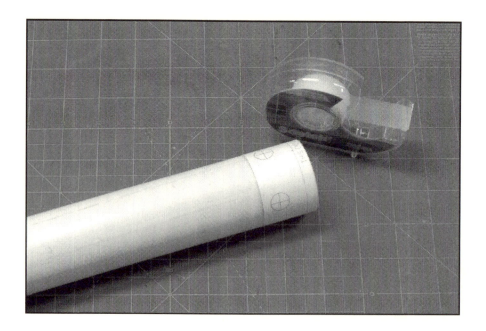

Step 07
Tape the 1 1/2" *PVC Nozzle insert template* onto a length of 1 1/2" schedule 40 PVC pipe.

Materials
- 1 1/2" Schedule 40 PVC pipe
- Scotch tape
- 1 1/2" PVC Nozzle insert template

Tools
- None

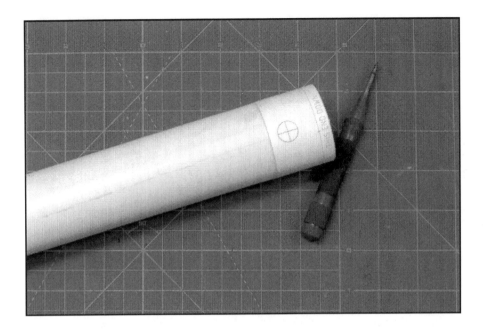

Step 08
Mark the red line on the template with a center punch or awl. This punch mark will be used as a guide for how far to insert the nozzle insert into the 1 1/2" x 2" PVC bushing in step 13.
The center punch will transfer a small punched hole to the pipe but the same job can be accomplished with a pen or a sharp nail.

Materials
- 1 1/2" Schedule 40 PVC pipe
- 1 1/2" PVC Nozzle insert template

Tools
- Center punch (McMaster part number 3489A13)

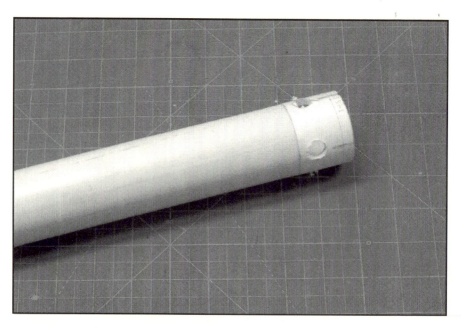

EYE PROTECTION
REQUIRED

KEEP HANDS
CLEAR

Step 09

Drill the 1/2" holes on the nozzle insert using the template as a guide. The 1/2" drill bit listed below is a Forstner drill bit which cuts the PVC very easily and cleanly. This type of drill bit is more expensive than a traditional bit and not as widely available. Either type can be used with equal success.

Materials
- 1 1/2" Schedule 40 PVC pipe
- 1 1/2" PVC Nozzle insert template

Tools
- 1/2" Forstner drill bit (McMaster part number 3216A22)
- Drill press or hand held drill

CAUTION EYE PROTECTION REQUIRED | CAUTION EAR PROTECTION REQUIRED | CAUTION KEEP HANDS CLEAR

Step 10
Cut the nozzle insert to size using the template as a guide.

Materials
- None

Tools
- Electric miter saw or PVC pipe cutter

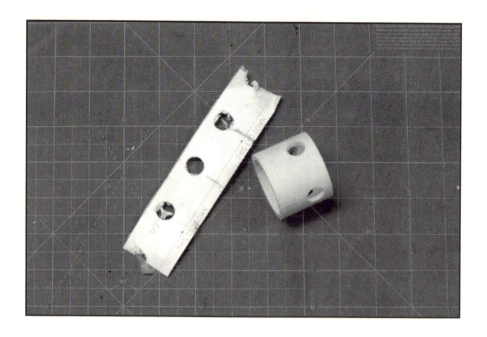

Step 11
Remove the template and set the nozzle insert aside.

Materials
- None

Tools
- None

EYE PROTECTION
REQUIRED

Step 12
Glue a 1 1/2" x 2" PVC reducing bushing into a 2" PVC coupling according to the instructions on the PVC glue you are using. I use Christy's Red Hot Blue Glue PVC Cement which does not require a primer.

Materials
- Christy's Red Hot Blue Glue PVC Cement or similar PVC cement with primer
- 2" Schedule 40 PVC Coupling
- 1 1/2" x 2" Schedule 40 PVC reducing bushing

Tools
- None

EYE PROTECTION
REQUIRED

Step 13
Glue the nozzle insert from step 11 into the 1 1/2" x 2" PVC bushing according to the instructions on the PVC glue you are using. I use Christy's Red Hot Blue Glue PVC Cement which does not require a primer. **Only insert the nozzle insert into the bushing up to the mark you made with the center punch in step 8.**

Materials
- Christy's Red Hot Blue Glue PVC Cement or similar PVC cement with primer
- Nozzle insert

Tools
- None

OPTIONAL
Let the PVC glue dry according to the manufacturer's instructions. Cut out the 2" *PVC Coupling retaining bolt template* from the template sheet. This optional step will make four 1/4-20 threaded holes in the nozzle. These threaded holes will allow the engine to be bolted inside a rocket. This step is not necessary if you are going to static test the engine or have an alternate method of securing the engine to a rocket body.

Materials
- 2" PVC Coupling retaining bolt template

Tools
- None

OPTIONAL

Tape the 2" *PVC Coupling retaining bolt template* around the 2" PVC coupling on the same side as the 1 1/2" PVC nozzle insert. Line up the template so that the holes you will be drilling will be **in-between** the 1/2" holes you already drilled on the 1 1/2" PVC nozzle insert.

Materials
- 2" PVC Coupling retaining bolt template
- Scotch tape

Tools
- None

EYE PROTECTION REQUIRED   **KEEP HANDS CLEAR**

OPTIONAL
Using a number 7 drill bit, drill the four holes marked on the 2" *PVC Coupling retaining bolt template* all the way through the 2" PVC coupling, the 1 1/2" x 2" PVC bushing and the 1 1/2" PVC nozzle insert.

Materials
- PVC Nozzle assembly

Tools
- #7 Drill bit for 1/4-20 tap (McMaster part number 2930A17)

CAUTION
EYE PROTECTION
REQUIRED

OPTIONAL

Remove the 2" *PVC Coupling retaining bolt template*. Insert a 1/4-20 tap into the chuck of a drill and tap all four holes. This can also be done by hand.

Note: If you need help with tapping, see *List of Links*.

Materials
- PVC Nozzle assembly

Tools
- 1/4-20 tap (McMaster part number 2521A571)
- Cordless drill

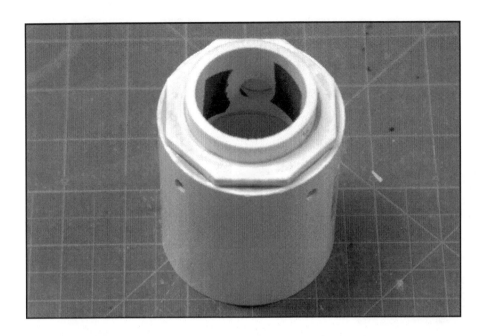

OPTIONAL
Cut four small pieces of masking tape and place them on the inside of the nozzle assembly over the four 1/4-20 holes. This will prevent the 1/4-20 holes from being filled with the hydraulic cement that will be used in the next several steps.

Materials
- PVC Nozzle assembly
- Masking tape

Tools
- None

Step 14

Put the entire nozzle assembly over the divergent paper cone from step 6. The paper cone should fit snuggly in the bottom of the 1 1/2" PVC nozzle insert.

Materials
- PVC Nozzle assembly
- Divergent paper cone

Tools
- None

EYE PROTECTION
REQUIRED

Step 15
Mix 3 ounces of hydraulic cement and 1.5 ounces of water in a Ziploc bag. Mix them together thoroughly until the cement is smooth and free of lumps. Hydraulic cement is different from traditional cement in that it expands as it dries to form a tight seal against whatever it is poured into. Common trade names for hydraulic cement are *Pourstone and Rocktite.*

Materials
- Hydraulic cement (McMaster part number 7685T42)
- 3 Ounce disposable cup
- Ziploc bag

Tools
- None

EYE PROTECTION
REQUIRED

Step 16
Cut a small corner off the edge of the Ziploc bag and pour the hydraulic cement into the nozzle assembly. Pour the cement in slowly so that air bubbles will not be trapped along the nozzle wall or in the 1/2" holes in the nozzle insert.

Materials
- None

Tools
- None

EYE PROTECTION
REQUIRED

Step 17
Once the hydraulic cement is about half the way up the divergent cone, put it aside and carefully drop in a 0.781" ID thick steel washer on the top of the divergent cone. If you have made the divergent cone correctly, the washer will rest about 3/8" below the top of the 1 1/2" x 2" PVC bushing.

Materials
- Thick steel washer 0.781" ID, 1.625" OD, 0.17" thick. (McMaster part number 98029A036)

Tools
- None

Step 18
Gently push down on the steel washer with your fingers until it is level and firm on the cone.

Materials
- None

Tools
- None

EYE PROTECTION
REQUIRED

Step 19
Cover the steel washer with hydraulic cement. Continue to cover the steel washer until the cement has covered the bushing and pipe stop inside the 2" PVC coupling.

Materials
 • None

Tools
 • None

Step 20
Allow the hydraulic cement to dry at least 45 minutes.

Materials
- None

Tools
- None

Step 21
Once the hydraulic cement is dry, use a pair of needle nose pliers to pull the divergent cone out of the nozzle. It should come out cleanly and easily. Leave the nozzle on its side to dry overnight.

Materials
• None

Tools
• Needle nose pliers

CAUTION | CAUTION
EYE PROTECTION | KEEP HANDS
REQUIRED | CLEAR

Step 22
Once the nozzle is dry, use a 49/64" drill bit to drill out the throat of the nozzle. This bit will just fit inside the steel washer which is now incased in hydraulic cement.
The nozzle is now completed.

Materials
- None

Tools
- 49/64" Drill bit. (McMaster part number 2933A59)
- Drill press or cordless drill

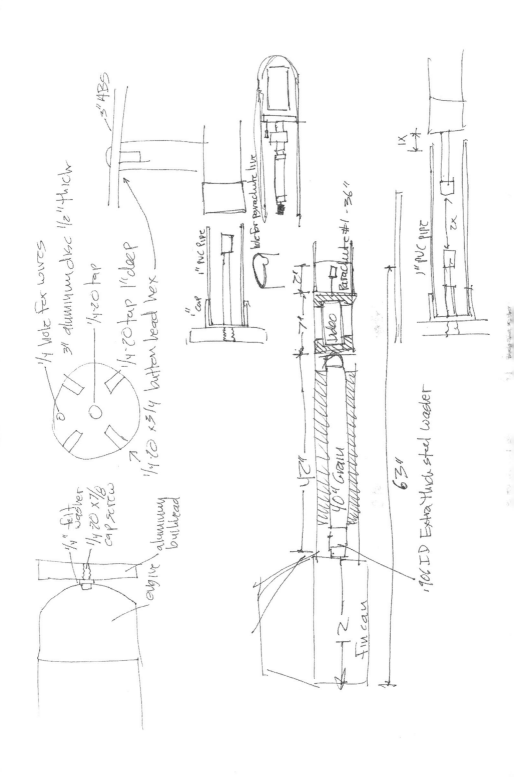

¼ hole for wires

3" aluminum disc ½" thick

¼-20 tap

¼-20 tap 1" deep

¼-20 x ¾ button head hex

¼" felt washer
¼-20 x ⅞ cap screw

engine aluminum bulkhead

3" ABS

Hole for parachute line

" cap   1" PVC pipe

video

Parachute #1 - 36"

7"   2"

4"

40" Grain

63"

tin can

.906 ID Extra thick steel washer

1X

2X

1" PVC pipe

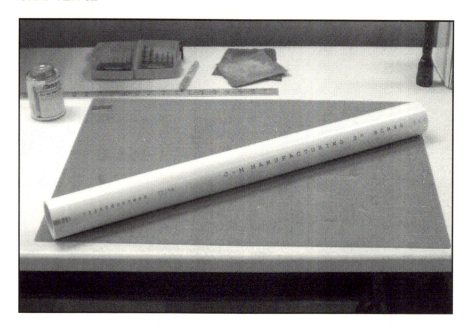

Chapter 02

CUTTING THE ENGINE CASING

The engine casing needs to hold in all the pressure produced by the burning propellant. Since the structural integrity of PVC degrades rapidly with an increase in temperature, the fuel burns from the inside out, never touching the casing until the propellant has finished burning.

CUTTING THE ENGINE CASING

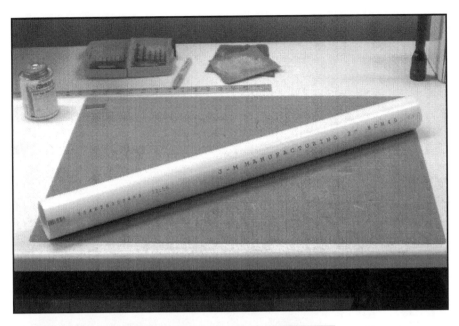

CAUTION	CAUTION	CAUTION
EYE PROTECTION REQUIRED	EAR PROTECTION REQUIRED	KEEP HANDS CLEAR

Step 01

The K450 engine uses a 23" case bonded fuel grain. This means the fuel is cast in the engine casing and is bonded to the sides of the PVC pipe. In order to allow room for the igniter and the hot fuel which will shrink as it cools, we need to make the initial casing length a little longer than the fuel grain. Cut a length of 2" PVC pipe 27" long.

Materials
- 2" Schedule 40 PVC Pipe

Tools
- Electric miter saw or PVC pipe cutter

**EYE PROTECTION
REQUIRED**

Step 01
Glue the 27" PVC engine casing into the finished nozzle assembly
according to the instructions on the PVC glue you are using. I use Christy's
Red Hot Blue Glue PVC Cement which does not require a primer.
The engine casing is now completed.

Materials
- Christy's Red Hot Blue Glue PVC Cement or similar PVC cement
  with primer
- Finished nozzle assembly

Tools
- Electric miter saw or PVC pipe cutter

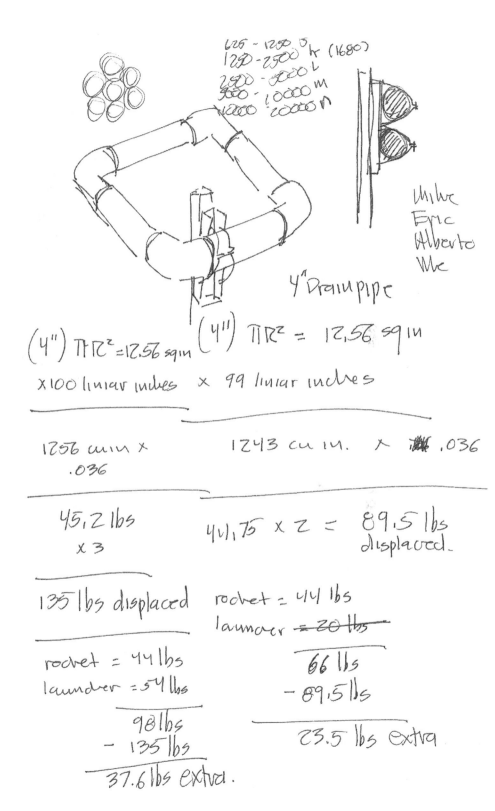

625 - 1250 s
1250 - 2500 k (1680)
2500 - 5000 L
5000 - 10000 m
10000 - 20000 n

Mike
Eric
Alberto
Me

4" Drainpipe

$(4") \pi R^2 = 12.56$ sq in $(4") \pi R^2 = 12.56$ sq in

x 100 liniar inches   x 99 liniar inches

1256 cu in x   1243 cu in.   x   .036
.036

45.2 lbs   44.75 x 2 = 89.5 lbs
x 3   displaced.

135 lbs displaced   rocket = 44 lbs
                         launcher = 20 lbs

rocket = 44 lbs   66 lbs
launcher = 54 lbs   - 89.5 lbs

98 lbs   23.5 lbs extra
- 135 lbs

37.6 lbs extra.

Chapter 03

MAKING THE IGNITER

The engine igniter is a crucial part in the proper operation of the K450 engine. The igniter contains a powered mix of sugar and Kno3 (Potassium Nitrate). When ignited by a small length of nichrome wire, the igniter produces a hot flame that fills the core of the engine, igniting all the exposed fuel at the same time.

EYE PROTECTION
REQUIRED

Step 01
A good digital scale is invaluable when trying to achieve consistent results. The model specified below is expensive but no other tool is as important when mixing fuel. A good digital scale that is accurate to 1/100th of a pound will not only allow you to make the K450 engine correctly but will allow you to experiment on your own for years to come. Measure out 1/2 pound of Kno3 (Potassium Nitrate).
For more information on Kno3 see *About Potassium Nitrate*.

Materials
- Plastic container
- Kno3 (Potassium Nitrate)

Tools
- Digital scale with accuracy to 1/100th of a pound. (McMaster part number 18605T86)

Step 02

Place the Kno3 (Potassium Nitrate) on an 8" x 8" metal pan and heat it in the toaster oven at 300 degrees Fahrenheit for 30 minutes to remove any moisture.

Materials
- 8" x 8" Metal pan
- Kno3 (Potassium Nitrate)

Tools
- Toaster oven

Step 03
Remove the Kno3 (Potassium Nitrate) after it is done heating and let it cool for 10 minutes. Make sure to wear oven mitts or heavy duty work gloves when removing the metal pan from the toaster oven as it will be very hot.

Materials
- 8" x 8" Metal pan
- Kno3 (Potassium Nitrate)

Tools
- Heavy duty work gloves

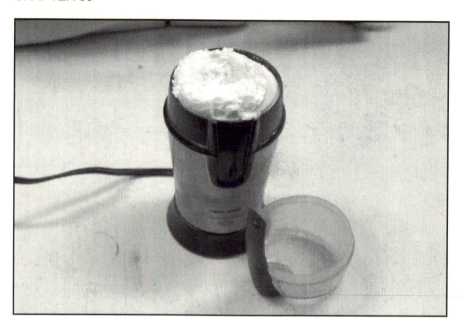

EYE PROTECTION
REQUIRED

Step 04
Grind the Kno3 (Potassium Nitrate) in a coffee grinder until the particle size has been sufficiently reduced to a fine powder.

Materials
- None

Tools
- Coffee grinder

CAUTION
EYE PROTECTION
REQUIRED

Step 05
Measure 0.26 lbs. of the dried and ground Kno3 (Potassium Nitrate) from the previous steps on the digital scale and put it into a plastic container with a lid.

Materials
- Plastic container with a lid

Tools
- Digital scale with accuracy to $1/100^{th}$ of a pound. (McMaster part number 18605T86)

DANGER

FLAMMABLE

Step 06

Measure 0.14 lbs. of powdered sugar on the digital scale and put it into the plastic container along with the Kno3.

For more information on powdered sugar see *About Powdered Sugar*.

Materials
- Plastic container with a lid

Tools
- Digital scale with accuracy to 1/100th of a pound. (McMaster part number 18605T86)

Step 07
Thoroughly mix the Kno3 (Potassium Nitrate) and powdered sugar together with a wooden spoon. This will yield 0.4 lbs. of igniter mixture which is 65% kno3 (Potassium Nitrate) and 35% sugar. Following this ratio, a larger or smaller amount of igniter mixture can be made at any time.

IMPORTANT NOTE: At this point the igniter mixture is flammable and should be treated accordingly. Store the igniter mixture away from heat and sources of ignition.

Materials
• None

Tools
• Wooden spoon

FLAMMABLE

Step 08

Seal the plastic container with a lid and set it aside. This mixture should always be kept tightly sealed since it absorbs moisture from the air which can affect its performance.

Materials
- Plastic container with lid

Tools
- None

CAUTION	CAUTION	CAUTION
EYE PROTECTION REQUIRED	EAR PROTECTION REQUIRED	KEEP HANDS CLEAR

Step 09
Cut a piece of 1" Schedule 40 PVC pipe 1" long.

Materials
- 1" Schedule 40 PVC pipe

Tools
- Electric miter saw or PVC pipe cutter

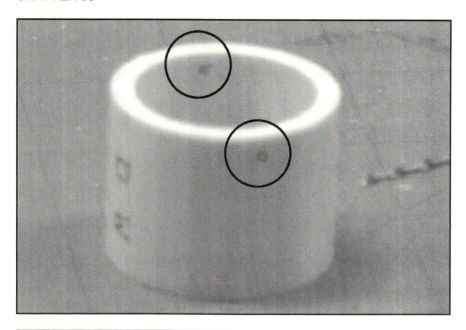

CAUTION	CAUTION
EYE PROTECTION REQUIRED	KEEP HANDS CLEAR

Step 10

Drill a 1/16" hole near the top of the 1" PVC pipe and out through the other side and set the 1" PVC aside.

Materials

- 1" Schedule 40 PVC pipe

Tools

- 1/16" Drill bit
- Cordless drill

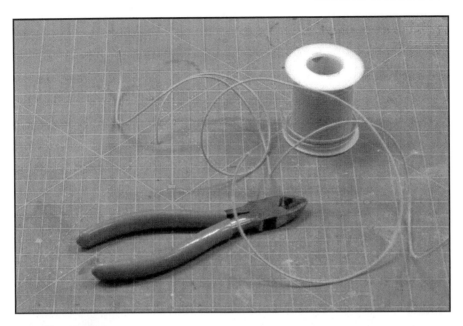

EYE PROTECTION
REQUIRED

Step 11
Cut two pieces of 22 gauge stranded wire 12" long.

Materials
- Stranded single conductor wire 22 AWG (McMaster part number 7587K931)

Tools
- Wire cutters

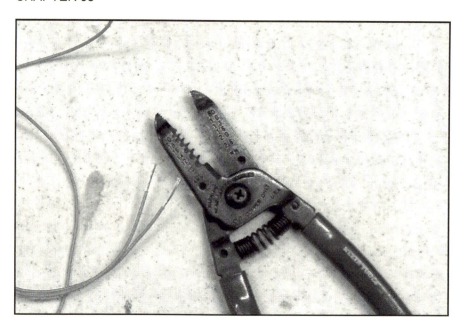

EYE PROTECTION
REQUIRED

Step 12
Strip the ends of the 22 gauge wires back 1/2".

Materials
- Stranded single conductor wire 22 AWG (McMaster part number 7587K931)

Tools
- Wire strippers

EYE PROTECTION
REQUIRED

Step 13
Cut a piece of 36 gauge nichrome wire about 6" long. The nichrome wire will be used to light the igniter mixture. It will heat up when voltage is applied to it, instantly lighting the igniter mixture which will immediately start the engine.

Materials
- 36 AWG Nichrome wire (McMaster part number 8880K85)

Tools
- Wire cutters

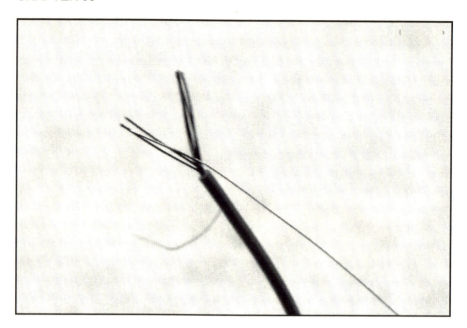

EYE PROTECTION
REQUIRED

Step 14

Split the strands on one section of 22 gauge wire and thread one end of the nichrome wire through it.

Materials
- None

Tools
- None

EYE PROTECTION
REQUIRED

Step 15
Wrap the nichrome wire around the strands below where they are split six times.

Materials
- None

Tools
- None

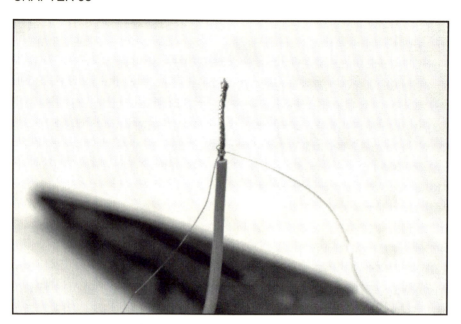

EYE PROTECTION
REQUIRED

Step 16
Close the strands and twist them together with a pair of needle nose pliers until they are tightly wound. This will trap the single strand of nichrome wire in the twisted strands of 22 gauge wire.

Materials
- None

Tools
- Needle nose pliers

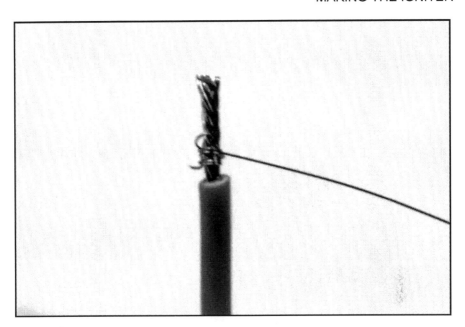

EYE PROTECTION
REQUIRED

Step 17
Trim the 22 gauge wire down to 1/4" and trim the excess nichrome wire with a pair of wire cutters.

Materials
- None

Tools
- Wire cutters

EYE PROTECTION
REQUIRED

Step 18
Thread one end of the nichrome wire through the 1/16" holes on the 1" PVC section from step 10.

Materials
- 1" PVC Pipe section

Tools
- None

EYE PROTECTION
REQUIRED

Step 19
Attach the loose end of nichrome wire to the second 22 gauge wire cut in
step 11 by repeating steps 14-17.

Materials
- None

Tools
- None

EYE PROTECTION
REQUIRED

Step 20
Twist the 22 gauge wire to roll up any slack in the nichrome wire so that the nichrome wire is taught across the 1" PVC section.

Materials
- None

Tools
- None

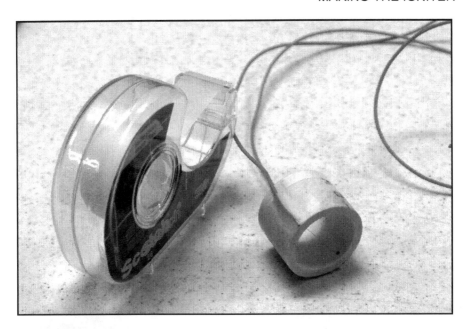

EYE PROTECTION
REQUIRED

Step 21
Wrap a piece of scotch tape around the 1" PVC section to secure the wires in place and set it aside.

Materials
- Scotch tape

Tools
- None

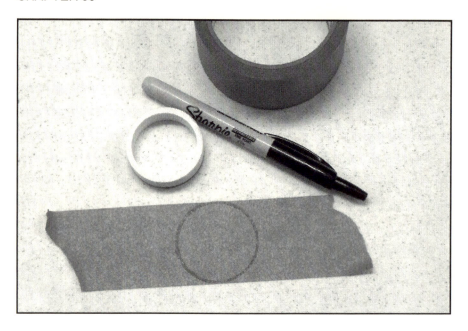

Step 22

Using the 1 1/2" PVC ring from Chapter 1, step 4, mark a circle on a piece of 2" wide masking tape using the outside of the PVC ring.

Materials
- 2" Wide masking tape
- Black marker

Tools
- None

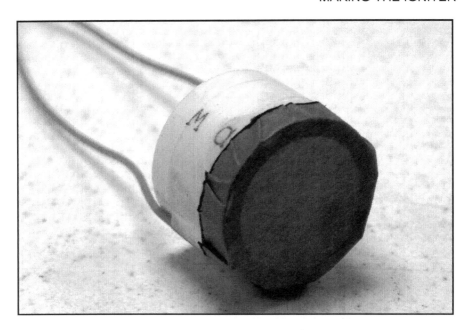

Step 23
Cut the circle out from the 2" masking tape and place it on the nichrome wire end of the 1" PVC pipe section. Fold the edges of the tape over the side of the 1" PVC pipe section and push them down to secure them to the pipe.

Materials
• 2" Wide masking tape

Tools
• Scissors

Step 24
Wrap a piece of 1" wide masking tape around the 1" PVC pipe section at least two times.

Materials
- 1" Wide masking tape

Tools
- Scissors

Step 25

Using a 1/4 teaspoon measuring spoon, start to add a small amount of igniter mixture from step 8 to the 1" PVC pipe section. Press the igniter mixture down gently over the nichrome wire until the powder is firm.

Materials
- Igniter mixture

Tools
- 1/4 Teaspoon measuring spoon

CAUTION **DANGER**
EYE PROTECTION
REQUIRED **FLAMMABLE**

Step 26
Continue adding igniter mixture to the 1" PVC pipe section and gently packing it down until the pipe section is full. Use the 1/4 teaspoon measuring spoon to make a small concave indentation in the igniter mixture at the top of the pipe section. This will be used to hold the epoxy in the next step.

Materials
- Igniter mixture

Tools
- 1/4 Teaspoon measuring spoon

CAUTION DANGER
EYE PROTECTION REQUIRED FLAMMABLE

Step 27

Clean off any loose igniter mixture from the edge of the pipe section. Mix a small amount of two part 5 minute epoxy and pour it evenly over the top of the igniter mixture ensuring that all the igniter mixture has been covered as well as the top face of the 1" PVC pipe section. Let the epoxy dry according to the manufacturer's instructions.

Materials
- Two part 5 minute epoxy (McMaster part number 7670A22)
- Mixing cup and disposable stirrer

Tools
- None

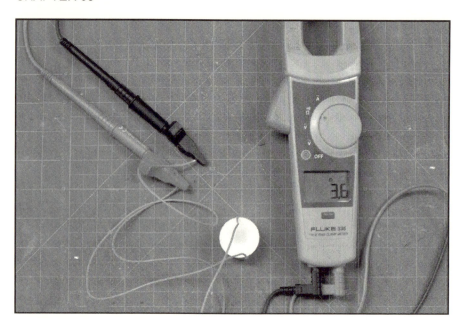

FLAMMABLE

Step 28
Check the igniter for continuity with a continuity tester or a multi-meter with Ohms scale. A reading of 3 to 5 Ohms is normal.

Materials
- None

Tools
- Multi-meter with Ohms scale

Step 29
Put a small zip tie around the wires at the top of the igniter to hold them securely in place. Trim the zip tie. The igniter is now completed

Materials
- Small zip ties (McMaster part number 7130K42)

Tools
- Wire cutters

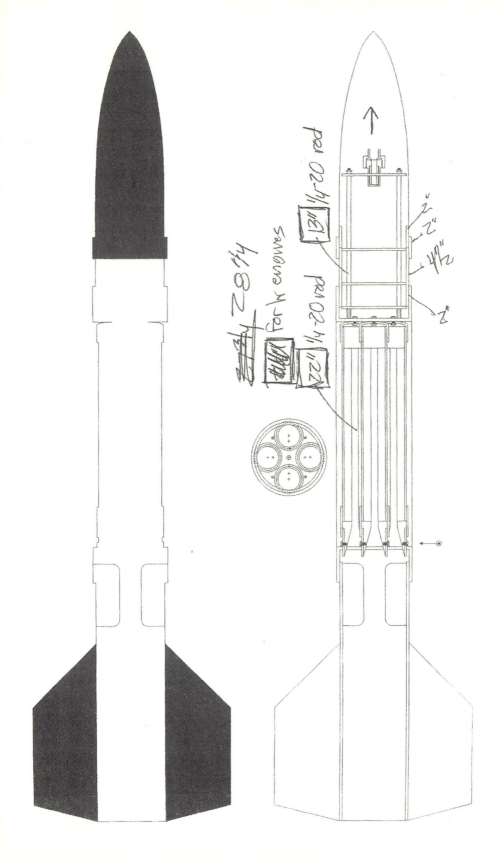

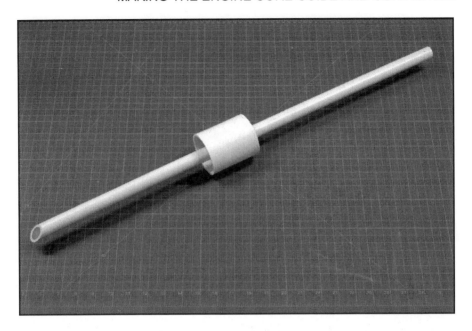

Chapter 04

MAKING THE ENGINE CORE GUIDE AND CORING ROD

When casting the fuel in the engine casing, a hollow cylinder needs to be cast in the center of the fuel from the nozzle to the igniter. The coring rod will perform this function and the core guide will ensure that the coring rod forms a cylinder in the exact center of the fuel. Once the fuel has solidified, the coring rod and core guide can be removed from the engine and used again and again.

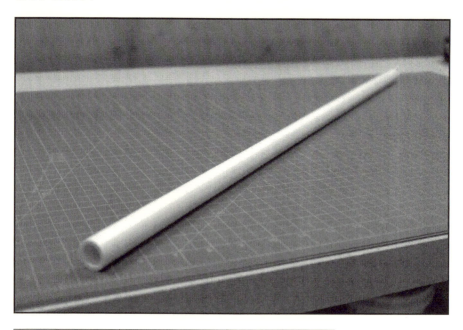

CAUTION	**CAUTION**	**CAUTION**
EYE PROTECTION REQUIRED	EAR PROTECTION REQUIRED	KEEP HANDS CLEAR

Step 01

Cut a 38" length of 1/2" Schedule 40 PVC pipe. This pipe will become the coring rod.

Materials

- 1/2" Schedule 40 PVC Pipe

Tools

- Electric miter saw or PVC pipe cutter

Step 02
Cover one end of the 1/2" PVC pipe with masking tape.

Materials
- Masking tape

Tools
- None

EYE PROTECTION
REQUIRED

Step 03
Mix 12 ounces of hydraulic cement and 6 ounces of water in a quart size plastic container. Mix them together thoroughly until the cement is smooth and free of lumps. Hydraulic cement is different from traditional cement in that it expands as it dries to form a tight seal against whatever it is poured into. Common trade names for hydraulic cement are *Pourstone and Rocktite.*

Materials
- Hydraulic cement (McMaster part number 7685T42)
- Plastic quart container

Tools
- None

EYE PROTECTION REQUIRED

Step 04
Secure the 1/2" PVC pipe to a vertical surface with the taped end on the bottom. Fill the pipe with hydraulic cement and let it harden for 45 minutes.

Materials
- Hydraulic cement (McMaster part number 7685T42)

Tools
- Clamps or similar device to secure the pipe vertically

CAUTION	CAUTION	CAUTION
EYE PROTECTION REQUIRED	EAR PROTECTION REQUIRED	KEEP HANDS CLEAR

Step 05
Once the hydraulic cement in the 1/2" PVC pipe has hardened, remove the masking tape from the bottom end. Cut the bottom end off the 1/2" PVC pipe at a 45 degree angle using an electric miter saw or hacksaw.

Materials
- None

Tools
- Electric miter saw or hacksaw

CAUTION	CAUTION	CAUTION
EYE PROTECTION REQUIRED	EAR PROTECTION REQUIRED	KEEP HANDS CLEAR

Step 06
Cut the top end of the 1/2" PVC pipe off so that the remaining piece is 36" long. Put the coring rod aside for now.

Materials
- None

Tools
- Electric miter saw or hacksaw

CAUTION
EYE PROTECTION
REQUIRED

Step 07
Prime and glue a 2" x 1" PVC reducing bushing into one end of a 2" PVC coupling according to the instructions on the PVC glue you are using. I use Christy's Red Hot Blue Glue PVC Cement which does not require a primer.

Materials
- Christy's Red Hot Blue Glue PVC Cement or similar PVC cement with primer
- 2" PVC Coupling
- 2" x 1" PVC Reducing bushing

Tools
- None

CAUTION
EYE PROTECTION
REQUIRED

Step 08
Prime and glue a 1" x 1/2" PVC reducing bushing into the 2" x 1" PVC reducing bushing according to the instructions on the PVC glue you are using. I use Christy's Red Hot Blue Glue PVC Cement which does not require a primer.

Materials
- Christy's Red Hot Blue Glue PVC Cement or similar PVC cement with primer
- 1" x 1/2" PVC Reducing bushing

Tools
- None

CAUTION **CAUTION**
EYE PROTECTION | KEEP HANDS
REQUIRED | CLEAR

Step 09
Using a 3/16" to 7/8" multi-diameter step drill bit, drill through the 1" x 1/2" PVC bushing so that the new bushing hole is 7/8" in diameter. The 1/2" PVC pipe that was assembled earlier should slip easily in and out of this hole but the hole should be tight enough so that there is very little play between the two pieces.

Note: A 7/8" spade bit can be used in place of the 3/16" to 7/8" step drill bit if a less expensive alternative is required.

Materials
• None

Tools
• Bench mounted drill press or cordless drill
• Multi-diameter step drill bit 3/16" to 7/8" (McMaster part number 89315A42)
• Bench vise

CAUTION | **CAUTION**
EYE PROTECTION REQUIRED | KEEP HANDS CLEAR

Step 10

Turn the 2" PVC coupling upside down so that the reducing bushings are on the bottom. Using a 2 3/8" hole saw, bore out the 2" PVC coupling down to the middle of the coupling. Boring out the coupling will allow the rod guide to be inserted onto the top of a 2" PVC engine casing and removed easily. The rod guide should slide easily on and off a piece of 2" PVC pipe but should be tight enough so that there is very little play between the two pieces.

Materials
- None

Tools
- 2 3/8" Hole saw (McMaster part number 4066A41 and 4066A79)
- Bench vise

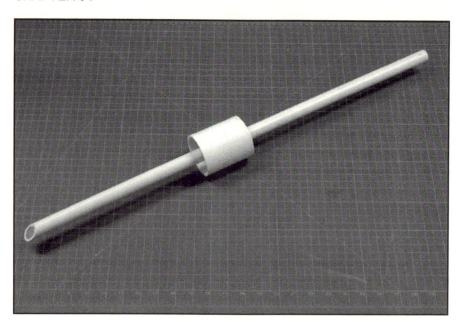

Step 11
The core guide and coring rod are now completed.

Materials
- None

Tools
- None

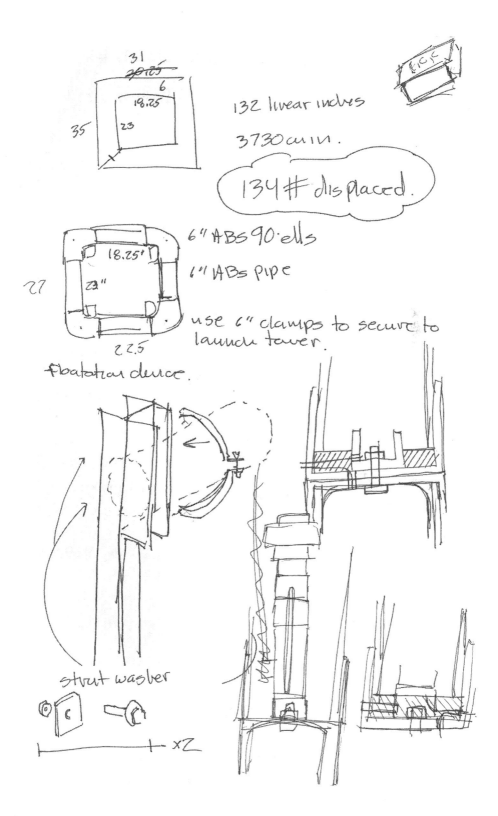

31
~~28.05~~
6
18.25
35
23

132 linear inches

3730 cu in.

134# displaced.

6" ABS 90·ells

1" ABS pipe

use 6" clamps to secure to launch tower.

18.25'
27
23"
22.5

flotation deuce.

strut washer

⊙ C

├──────┼── x2

Chapter 05

MAKING THE FUEL

The fuel for the K450 PVC engine is a precise mix of oxidizer and fuel. Both of these ingredients come together very quickly to create a safe, reliable fuel that takes only minutes to make.

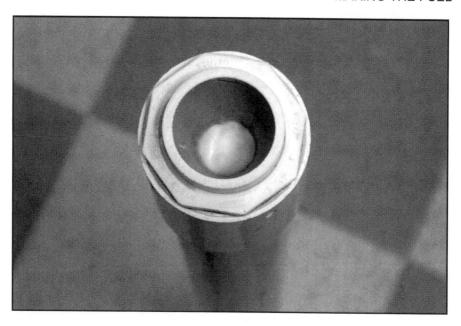

Step 01
Insert a cotton ball into the throat of the nozzle.

Materials
- Cotton balls

Tools
- None

Step 02
Add several more cotton balls to the nozzle so that the divergent section of the nozzle is full.

Materials
- Cotton balls

Tools
- None

Step 03
Using two pieces of 2" masking tape cover the nozzle opening and secure the cotton balls.

Materials
- 2" Masking tape

Tools
- None

Step 04
Cut a piece of 2" Schedule 40 PVC pipe 1/2" long.

Materials
- 2" Schedule 40 PVC Pipe

Tools
- Electric miter saw or pipe cutter

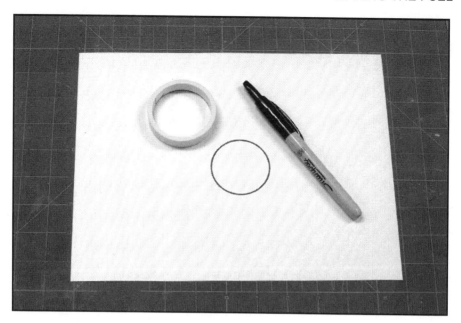

Step 05
Using the 2" x 1/2" PVC ring as a guide, mark the inside circumference on a piece of regular paper with a black marker.

Materials
- 8.5" x 11" Paper
- Black marker

Tools
- None

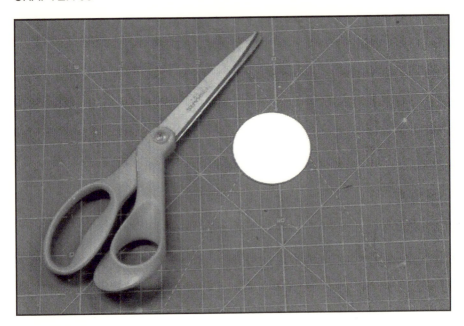

Step 06
Cut out the 2" diameter paper circle with a pair of scissors.

Materials
- None

Tools
- Scissors

Step 07
Drop the 2" diameter paper circle in the top of the engine casing letting it fall down to the nozzle throat. The paper circle will prevent fuel from flowing into the nozzle when the fuel is poured into the engine casing.

Materials
- Engine casing

Tools
- None

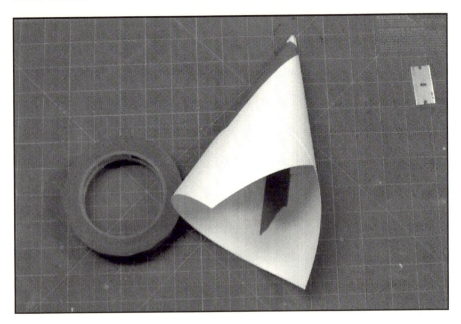

Step 08
Roll a sheet of 110 lb. card stock or similar heavy weight paper into a cone and tape the outside and inside seam with 1" masking tape.

Materials
- 110 lb. Card stock or similar heavy weight paper
- 1" Masking tape

Tools
- None

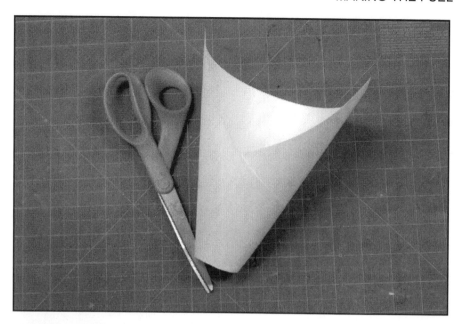

CAUTION
EYE PROTECTION
REQUIRED

Step 09
Cut a section off the tip of the cone so that it will just fit inside the engine casing.

Materials
- None

Tools
- Scissors

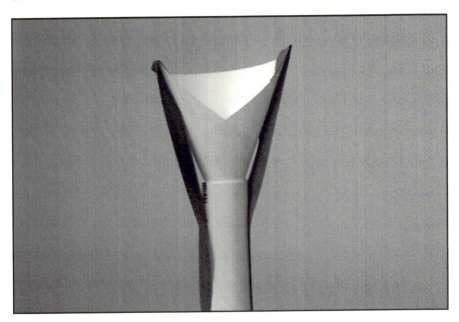

Step 10
Tape either side of the cone with 2" masking tape to the engine casing so it is secure.

Materials
- 2" Masking tape

Tools
- None

OPTIONAL
A wide mouth plastic funnel put in the top of the engine casing will serve the same purpose as the paper funnel. The plastic funnel can be easily cleaned and used again and again. Either one will work fine.

Materials
- None

Tools
- Wide mouth plastic funnel (McMaster part number 9017T15)

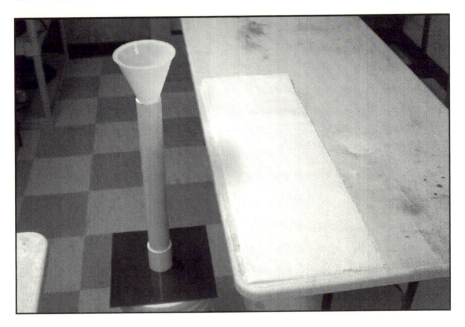

Step 11
Place a row of paper towels, 36" long on the work surface.

Materials
- Paper towels

Tools
- None

Step 12
Place an electric wok on the work surface. The wok will be used to heat the fuel and lower its viscosity so that it will pour easily into the engine casing.

Materials
- None

Tools
- Electric wok (Target part number 10771140)

EYE PROTECTION
REQUIRED

Step 13
Place the completed coring rod from Chapter 4 on the paper towels and generously spray it with WD-40 lubricant on all sides.

Materials
- WD-40 Lubricant
- Completed core guide and coring rod

Tools
- None

Step 14
Place the completed core guide from Chapter 4 on the paper towels next to the lubricated coring rod.

Materials
- None
- Completed core guide and coring rod

Tools
- None

Step 15
Place a pair of heavy work gloves and safety goggles next to the electric wok on the work surface.

Materials
- None

Tools
- Heavy work gloves
- Safety goggles

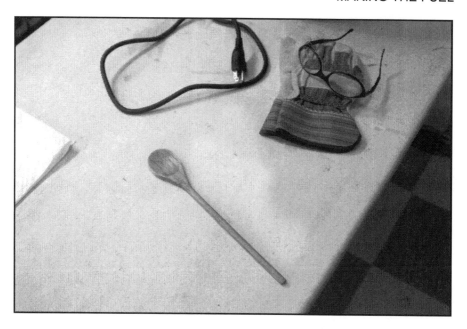

Step 16
Place a wooden spoon next to the work gloves on the work surface.

Materials
- None

Tools
- Wooden spoon

Step 17
Affix a single use temperature indicating label on the wooden spoon. The temperature dot will change color to black when the proper temperature for the fuel has been reached. The labels are an inexpensive way to determine when the fuel is at the proper temperature. An infrared thermometer or candy thermometer can be used as well.

Materials
- None

Tools
- Temperature indicating dots (McMaster part number 5952K223)
- Wooden spoon

EYE PROTECTION
REQUIRED

Step 18
Using a digital scale or some other accurate means of measurement, measure out 3.25 pounds of Kno3 (Potassium Nitrate) into a plastic quart container. For more information on Kno3 see *About Potassium Nitrate*.

Materials
- Plastic quart container
- Kno3 (Potassium Nitrate)

Tools
- Digital scale (McMaster part number 18605T86)

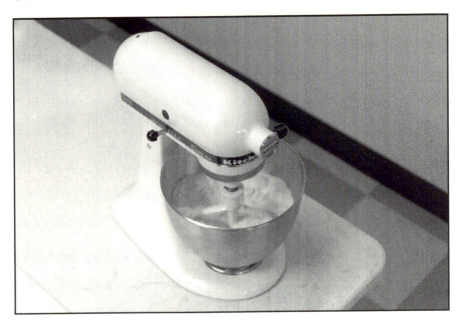

EYE PROTECTION
REQUIRED

Step 19
Put the 3.25 lbs. of Kno3 (Potassium Nitrate) in a stand mixer and mix on the lowest setting for 30 minuets. The Kitchen Aid stand mixer listed below is preferred but any electric stand mixer or hand mixer will work just fine. Even a potato masher will work if doing this step by hand and cost is a concern.

Materials
- Plastic quart container
- Kno3 (Potassium Nitrate)

Tools
- Electric stand mixer. The mixer used in this book is a Kitchen Aid K45SS model mixer. www.kitchenaid.com

EYE PROTECTION
REQUIRED

Step 20
Remove the Kno3 from the stand mixer and put it in an 8" x 8" metal pan.

Materials
- 8" x 8" Metal pan

Tools
- None

CAUTION — EYE PROTECTION REQUIRED   CAUTION — KEEP HANDS CLEAR   DANGER — HOT

Step 21
Put the 8" x 8" metal pan in a toaster oven and heat the Kno3 (Potassium Nitrate) at 300 degrees Fahrenheit for 30 minuets.

Materials
- 8" x 8" Metal pan
- Kno3 (Potassium Nitrate)

Tools
- Toaster oven

Step 22
Remove the Kno3 (Potassium Nitrate) from the toaster oven and let it cool for 10 minutes. Make sure to wear oven mitts or heavy duty work gloves when removing the metal pan from the toaster oven as it will be very hot.

Materials
- 8" x 8" Metal pan
- Kno3 (Potassium Nitrate)

Tools
- Heavy duty work gloves

Step 23
Take the bowl from the stand mixer or any other suitable container and place it on the digital scale. Set the scale to 0.00.

Materials
- Stand mixer bowl or any other suitable container

Tools
- Digital scale (McMaster part number 18605T86)

EYE PROTECTION
REQUIRED

Step 24
Measure out 2.77 lbs. of Kno3 into the stand mixer bowl using the digital
scale.

Materials
- Stand mixer bowl or any other suitable container

Tools
- Digital scale (McMaster part number 18605T86)

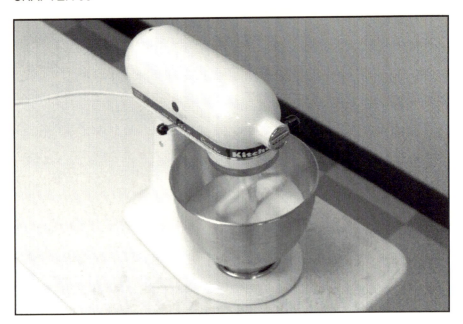

EYE PROTECTION
REQUIRED

Step 25
Put the stand mixer bowl with Kno3 (Potassium Nitrate) in it back on the mixer and start to mix it again on the lowest speed setting.

Materials
- Stand mixer bowl or any other suitable container

Tools
- Electric stand mixer. The mixer used in this book is a Kitchen Aid K45SS model mixer. www.kitchenaid.com

EYE PROTECTION
REQUIRED

Step 26
Using a digital scale or some other accurate means of measurement,
measure out 0.68 pounds of powdered sugar into a plastic quart container.
For more information on powdered sugar see *About Powdered Sugar*.

Materials
- Plastic quart container
- Powdered sugar

Tools
- Digital scale (McMaster part number 18605T86)

CAUTION
EYE PROTECTION
REQUIRED

Step 27
Stop the stand mixer and add the powdered sugar to the Kno3 (Potassium Nitrate) already in the bowl. Turn the stand mixer back on and mix the Kno3 and powdered sugar together while you complete the next two steps.

Materials
- Plastic quart container
- Powdered sugar

Tools
- Digital scale (McMaster part number 18605T86)

EYE PROTECTION
REQUIRED

Step 28
Place the electric wok on the digital scale and measure out 0.66 pounds of corn syrup into the electric wok. For more information on corn syrup see *About Corn Syrup*.

Materials
- Plastic quart container
- Corn syrup

Tools
- Digital scale (McMaster part number 18605T86)
- Electric wok (Target part number 10771140)

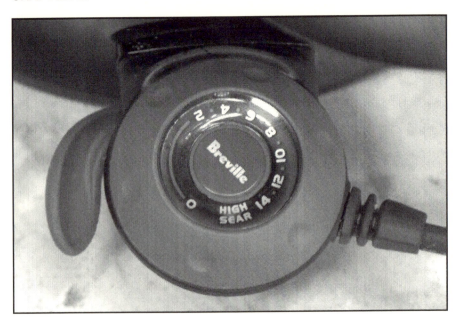

CAUTION
EYE PROTECTION
REQUIRED

DANGER
HOT

Step 29

Remove the electric wok from the digital scale and return it to the work surface. Turn the thermostat on the electric wok to its highest setting.

Materials
- None

Tools
- None

Step 30

Turn off the stand mixer and remove the mixing bowl containing the Kno3 (Potassium Nitrate) and powdered sugar mixture. Place the mixing bowl near the electric wok.

Materials
- None

Tools
- None

Step 31
Let the corn syrup heat up for about 60 seconds until small bubbles start to form near the surface. The time on this will vary depending on the electric wok wattage. Do not allow the corn syrup to boil.

Materials
  • None

Tools
  • None

CAUTION — EYE PROTECTION REQUIRED | CAUTION — KEEP HANDS CLEAR | DANGER — HOT | DANGER — FLAMMABLE

Step 32

Add all the Kno3 (Potassium Nitrate) and powdered sugar mixture into the wok and start to carefully stir it into the hot corn syrup. Be careful, the powdered sugar is very fine and will be forced out of the wok by rapid movements of the wooden spoon.

Materials
- None

Tools
- None

Step 33
Continue to stir the mixture until it starts to become fluid, about 2 minutes.

Materials
- None

Tools
- None

Step 34

After a few more minutes the mixture will become very fluid and much easier to stir. Continue to stir the mixture.

Materials
- None

Tools
- None

CAUTION — EYE PROTECTION REQUIRED

CAUTION — KEEP HANDS CLEAR

DANGER — HOT

DANGER — FLAMMABLE

Step 35
After a few more minutes the mixture will become very fluid and will be nearing the setpoint temperature of 210 degrees Fahrenheit. Continue to stir the mixture while periodically checking the temperature indicating dot on the wooden spoon.

Materials
- None

Tools
- None

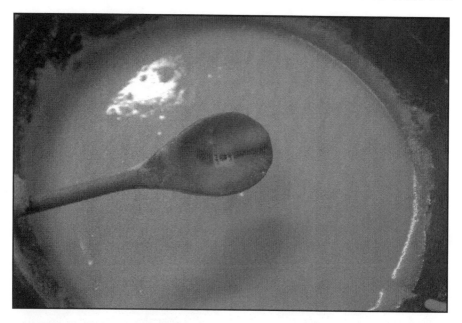

Step 36
Once the mixture reaches 210 degrees Fahrenheit, the dot on the temperature indicating label will turn black.

Materials
- None

Tools
- None

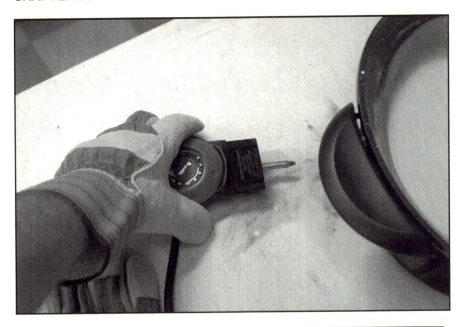

Step 37
Remove the wok thermostat so that no more heat will be introduced into the fuel.

Materials
- None

Tools
- None

CAUTION
EYE PROTECTION
REQUIRED

CAUTION
KEEP HANDS
CLEAR

DANGER
HOT

DANGER
FLAMMABLE

Step 38
Pour all the hot fuel into the engine casing using the paper or plastic funnel as a guide.

Materials
- None

Tools
- None

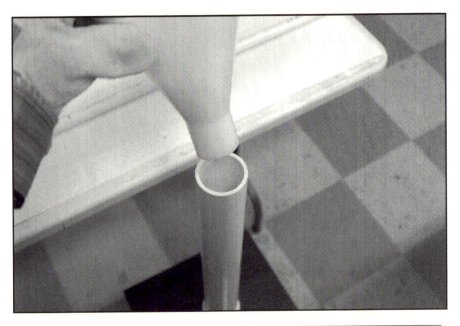

Step 39
Remove the paper or plastic funnel and set it aside on the work surface.

Materials
- None

Tools
- None

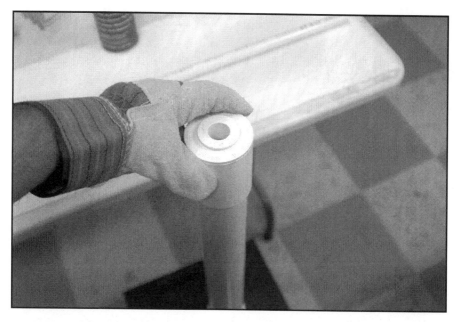

Step 40
Immediately place the completed rod guide from Chapter 4 on the top of the engine casing.

Materials
- None

Tools
- None

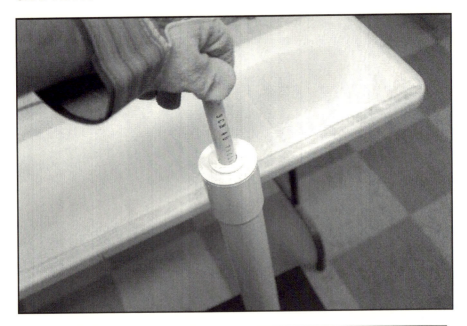

Step 41

Slide the lubricated coring rod through the rod guide and into the hot fuel. Slowly push the coring rod down until it reaches the bottom of the engine casing. Twist the coring rod until you feel it break the paper circle on the nozzle. Push the coring rod all the way through the nozzle until it rests on the bucket top.

Materials

- None

Tools

- None

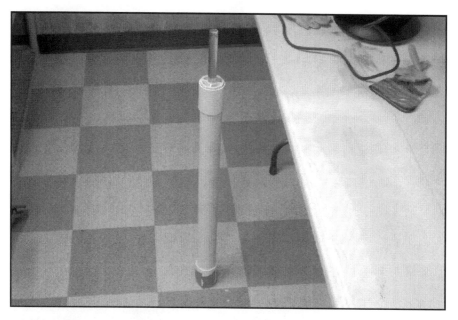

Step 42
Let the engine cool for 2 to 3 hours until the outside of the engine casing
has reached room temperature. Do not disturb the coring rod or core guide
during this time.

Materials
- None

Tools
- None

Step 43
Clean up the remaining fuel in the wok while it is still hot. Clean it under warm running water in a sink or with a hose. The fuel will dissolve in water and can safely be put down the drain. The fuel is now completed. After 2 to 3 hours of cooling time, proceed to the next chapter.

For reference and experimentation purposes, the K450 fuel consists of 65% Kno3 (Potassium Nitrate), 19% Corn Syrup and 16% Powdered Sugar.

Materials
- None

Tools
- None

wood.
gravel3 table
~~Powerstone~~
~~Phillip~~

5-3/4

5.75

6 1/8

7.5   cir.   33,16
      sq.   52.25

6 1/4 4

6.5 = 42,25

.000895
~~.000055~~
.00125
.00125

.34   16 x 24 Foam core
.18   4 x 36 Balsa wood
.09   3 x 24 Poster board

5.75

5.75

.048
.048
.14
.236

.045 .21 .045

.3 total

.048

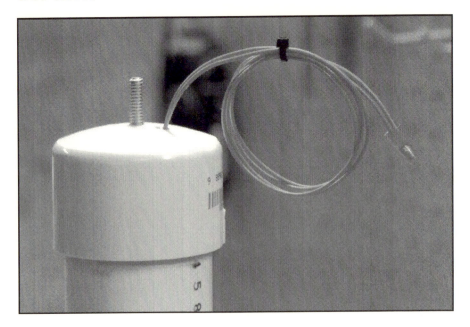

Chapter 06

INSTALLING THE END CAP AND IGNITER

The end cap holds in all the internal pressure of the engine as well as storing an ignition source for the fuel. If put together correctly, the end cap and the igniter should work without fail.

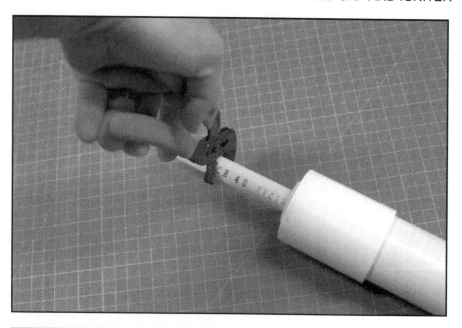

CAUTION
EYE PROTECTION
REQUIRED

DANGER
FLAMMABLE

Step 01
Grip the coring rod firmly with a pair of self-adjusting pliers.

Materials
- None

Tools
- Self-adjusting pliers (McMaster part number 51375A34)

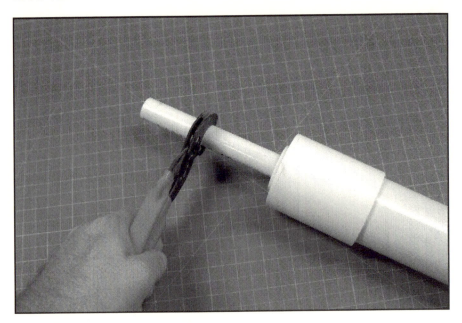

Step 02
Holding the engine casing in one hand and the self-adjusting pliers in another, twist the pliers downward 90 degrees.

Materials
- None

Tools
- Self-adjusting pliers (McMaster part number 51375A34)

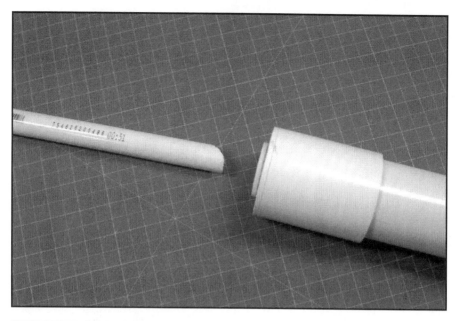

Step 03

While still holding the engine casing in one hand and the self-adjusting pliers in another, pull the coring rod out of the engine. The coring rod should slip easily out of the engine.

Materials

- None

Tools

- Self-adjusting pliers (McMaster part number 51375A34)

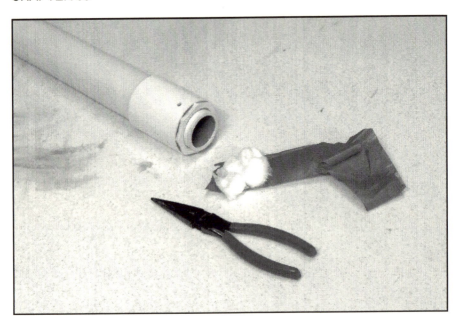

Step 04
Remove the 2" masking tape on the nozzle and the cotton balls inside. Needle nose pliers may be needed to remove the cotton ball in the throat of the nozzle because there will be a small amount of fuel on it.

Materials
- None

Tools
- Needle nose pliers

Step 05

Use a flashlight and inspect the fuel core that runs through the engine. The fuel should be smooth and even. There should not be any gaps or voids. If the fuel is less than perfect, soak the engine in a tub of warm water until the fuel has dissolved. Dispose of the engine casing and start the engine construction process again. Otherwise proceed with installing the end cap and igniter.

Materials
- None

Tools
- Flashlight

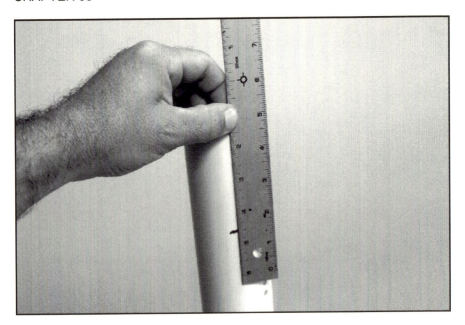

Step 06

Using a ruler, measure the distance inside the engine from the top of the fuel grain to the top of the engine casing. Mark the outside of the engine casing 1 1/2" above where the fuel ends.

Materials
- Black marker

Tools
- Ruler

Step 07

Cut the engine casing on the mark made in step 06. The engine casing should be 1 1/2" longer than the fuel grain.

Materials
- None

Tools
- Electric miter saw or pipe cutter

Step 08

Gently push a tapered polyethylene plug into the nozzle throat. Use a 3/8"
nut driver or similar blunt tool to press the plug into the throat and seal it.
The fuel for the K450 engine will absorb moisture if left exposed to the air
and will eventually become gummy and unusable. The polyethylene plug
will keep the engine dry and ready to use even after months of storage.

Materials
- Tapered polyethylene plug (McMaster part number 4491K16)

Tools
- 3/8" Nut driver

EYE PROTECTION
REQUIRED

Step 09
Mix a small amount of 5 minute two part epoxy in a small 3 ounce paper
cup.

Materials
- 5 Minute two part epoxy. (McMaster part number 7670A22)
- Mixing cup and disposable stirrer

Tools
- None

CAUTION
EYE PROTECTION
REQUIRED

DANGER
FLAMMABLE

Step 10
Using a flux brush or similar disposable brush; coat the top of the fuel with epoxy as well as the bottom edge of the igniter.

Materials
- 5 Minute two part epoxy. (McMaster part number 7670A22)
- Flux brush used for soldering copper pipe or similar disposable brush

Tools
- None

CAUTION
EYE PROTECTION
REQUIRED

DANGER
FLAMMABLE

Step 11

Install the igniter over the fuel core at the top of the fuel grain and press it
down gently so that there are no gaps between the igniter and the fuel core.

Materials
  • None

Tools
  • None

Step 12
Pour the remaining 5 minute epoxy around the edge where the igniter and the fuel meet.

Materials
- 5 Minute two part epoxy. (McMaster part number 7670A22)

Tools
- None

Step 13

Hold down the igniter until the epoxy has hardened and the igniter is firmly in place.

Materials
- None

Tools
- None

FLAMMABLE

Step 14
Mix two ounces of 5 minute epoxy in a small 3 ounce paper cup.

Note: To reduce the cost of the engine even further, any hydraulic cement can be used in this step as a substitute for 5 minute epoxy.

Materials
- 5 Minute two part epoxy. (McMaster part number 7670A22)
- Mixing cup and disposable stirrer
- Hydraulic cement (McMaster part number 7685T42)

Tools
- None

Step 15

Using a small piece of 1" masking tape, tape the igniter wires to the side of the engine casing so that they are centered at the top of the engine casing.

Materials
- 1" Masking tape

Tools
- None

Step 16
Pour the 5 minute epoxy (or hydraulic cement) over the igniter until it is flush with the top of the engine casing. Let the epoxy or cement dry according to the manufacturer's instructions.

.

Materials
- 5 Minute two part epoxy. (McMaster part number 7670A22)
- Hydraulic cement (McMaster part number 7685T42)

Tools
- None

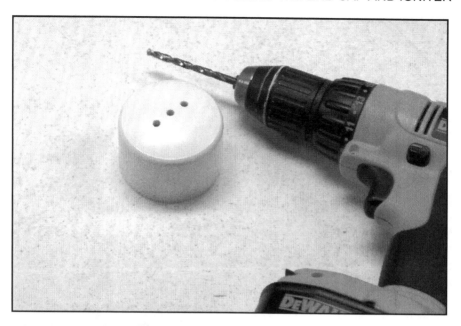

CAUTION | CAUTION
EYE PROTECTION REQUIRED | KEEP HANDS CLEAR

Step 17
Using a 1/4" drill bit, or a number 7 drill bit if you plan to install a 1/4-20 cap screw at the top of the engine for mounting purposes, drill three holes in a 2" PVC cap. Drill one in the center of the cap and the other two 1/4" to either side of the first hole.

Materials
- 2" Schedule 40 PVC Cap

Tools
- 1/4" Drill bit
- #7 drill bit for ¼-20 tap (Mcmaster part number 2930A17)
- Cordless drill or drill press

CAUTION | CAUTION
EYE PROTECTION | KEEP HANDS
REQUIRED | CLEAR

OPTIONAL
Insert a 1/4-20 tap into the chuck of a drill and tap the center hole. This can also be done by hand.

Note: If you need help with tapping, see *List of Links*.

Materials
- 2" Schedule 40 PVC Cap

Tools
- 1/4-20 Tap (McMaster part number 2521A571)
- Cordless drill or drill press

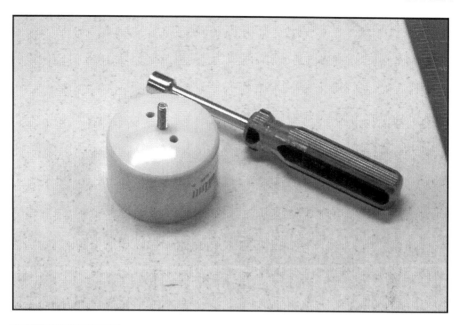

EYE PROTECTION
REQUIRED

OPTIONAL
Install a 1/4-20 x 1" cap screw through the 2" PVC cap and tighten with a
5/16" nut driver or adjustable wrench.

Materials
- 1/4-20 x 1" Cap screw (McMaster part number 92865A542)

Tools
- 5/16" Nut driver or adjustable wrench

143

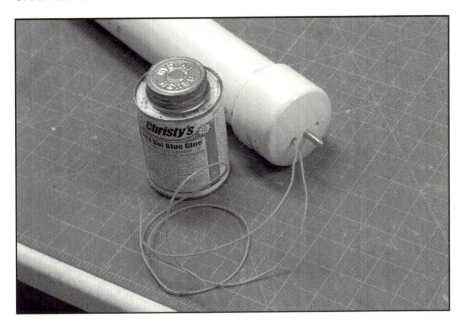

Step 18
Glue the 2" PVC cap to the top of the engine casing according to the instructions on the PVC glue you are using. I use Christy's Red Hot Blue Glue PVC Cement which does not require a primer. Before you glue the two pieces together, feed the two ignition wires through one of the holes in the 2" PVC cap.

Materials
- Christy's Red Hot Blue Glue PVC Cement or similar PVC cement with primer

Tools
- None

EYE PROTECTION
REQUIRED

Step 19
Mix 3 ounces of hydraulic cement and 1.5 ounces of water in a Ziploc bag. Mix them together thoroughly until the cement is smooth and free of lumps. Hydraulic cement is different from traditional cement in that it expands as it dries to form a tight seal against whatever it is poured into. Common trade names for hydraulic cement are *Pourstone and Rocktite.*

Materials
- Hydraulic cement (McMaster part number 7685T42)
- 3 Ounce disposable cup
- Ziploc bag

Tools
- None

EYE PROTECTION
REQUIRED

Step 20
Cut a small corner off the edge of the Ziploc bag and pour the cement back into the small 3 ounce paper cup.

Materials
- None

Tools
- None

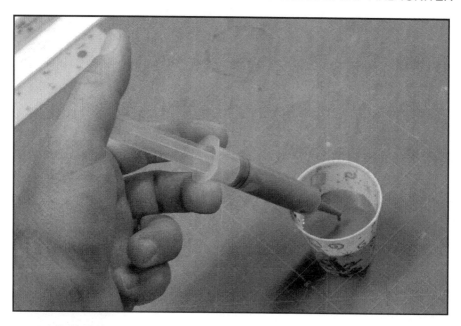

CAUTION
EYE PROTECTION
REQUIRED

Step 21
Purchase a small plastic syringe. The syringe will be used to inject hydraulic cement into the top of the engine cap to fill the void between the igniter and the cap. Pull up the hydraulic cement into the plastic syringe.

Materials
- Plastic syringe (McMaster part number 7510A662) Also available at most marine centers

Tools
- None

CAUTION **DANGER**
EYE PROTECTION
REQUIRED **FLAMMABLE**

Step 22
Inject the hydraulic cement into one of the two holes on top of the 2" PVC
cap until the cement comes out the other hole.

Materials
- Plastic syringe (McMaster part number 7510A662) Also available at
  most marine centers

Tools
- None

Step 23
Clean up any extra expanding cement that may have spilled out of the holes. Let the expanding cement dry according to the manufacturer's instructions.

Materials
- Paper towels

Tools
- None

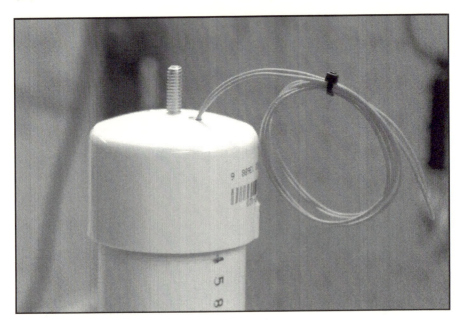

Step 24
Wrap up the ignition wires on top of the engine and secure them with a small zip tie.

Materials
- Small zip ties (McMaster part number 7130K42)

Tools
- None

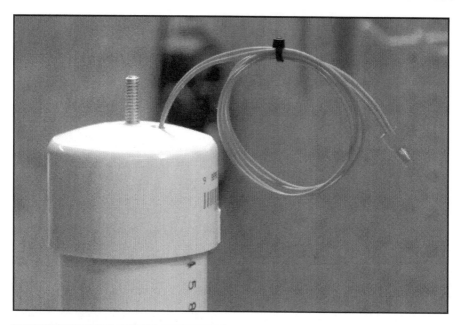

CAUTION
EYE PROTECTION
REQUIRED

DANGER
FLAMMABLE

Step 25
Shunt the two wires together with an insulated crimp on wire connector so
that there is no risk of accidental engine ignition. Congratulations, the K450
engine is now completed!

IMPORTANT: THE K450 ENGINE YOU HAVE JUST COMPLETED IS A
POWERFUL ENGINE THAT SHOULD BE HANDLED SAFELY AND
STORED SECURELY. NEVER, UNDER ANY CIRCUMSTANCES, PUT
POWER TO THE IGNITER LEADS WITH THE EXCEPTION OF
LAUNCHING OR TESTING THE ENGINE IN A REMOTE SAFE AREA.

Materials
- Small zip ties (McMaster part number 7130K42)

Tools
- None

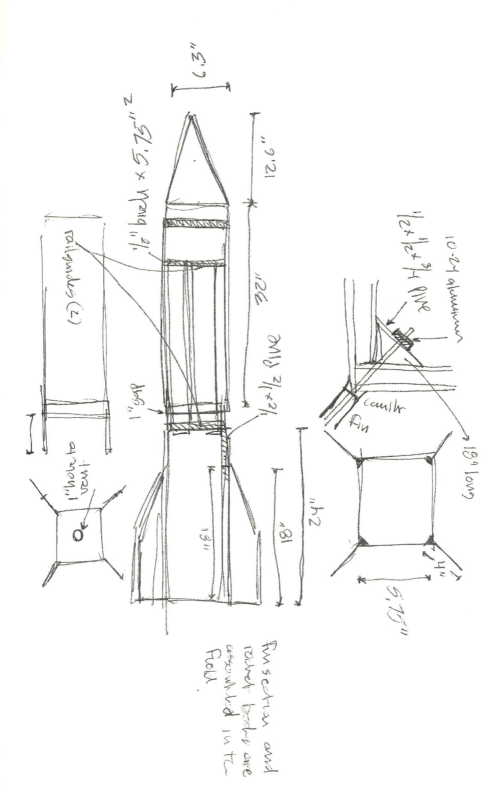

Chapter 07

TESTING AND LAUNCHING

Launching the K450 engine in a rocket body is a procedure that depends heavily on the design and construction of the rocket body itself. Since the construction of a rocket body is beyond the scope of this book, instructions for launching the K450 will not be covered here. Several mounting points have been worked into the design of the K450 which can be used to attach the K450 to a rocket body or completely omitted if necessary. The method listed below for static testing the K450 engine is one that has worked very well for the author but is by no means the only method available.

Step 01
Purchase two 2" conduit hangers and Install two washers, a nylon spacer, a 1/4-20 x 5/8" cap screw and nut on each one. See the picture above for clarification. The two washers will allow the conduit hangers and the engine to slide up and down a length of strut channel while testing.

Materials
- 1/4" x 1 1/4" Fender washer (McMaster part number 91090A109)
- Nylon unthreaded round spacer (McMaster part number 94639A570)
- 1/4-20 x 5/8" Cap screw (McMaster part number 92865A938)
- 1/4-20 Nut (McMaster part number 90494A029)

Tools
- 5/16" Nut driver or adjustable wrench

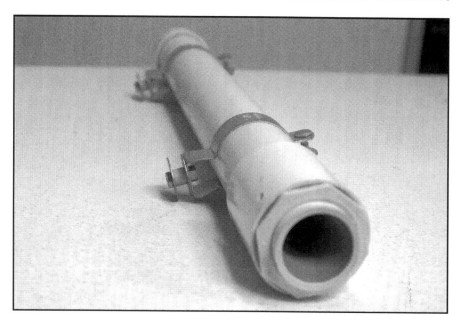

Step 02
Attach the two conduit hangers on the K450 engine in line with each other about 18" apart.

Materials
- 1/4" x 1 1/4" Fender washer (McMaster part number 91090A109)
- Nylon unthreaded round spacer (McMaster part number 94639A570)
- 1/4-20 x 5/8" Cap screw (McMaster part number 92865A938)
- 1/4-20 Nut (McMaster part number 90494A029)

Tools
- 5/16" Nut driver or adjustable wrench

Step 03
Purchase a hydraulic scale and a 0-400 PSI bottom connection gauge. Attach the 0-400 PSI gauge to the hydraulic scale.

Materials
- Hydraulic scale (McMaster part number 17645T21)
- 0-400 PSI 4 1/2" gauge (McMaster part number 3846K21)

Tools
- Adjustable wrench

EYE PROTECTION
REQUIRED

Step 04
Assemble a test stand using 1 5/8" x 1 5/8" strut channel.

Materials
- 1 5/8" x 1 5/8" steel strut channel (McMaster part number 3310T39)

Tools
- Adjustable wrench

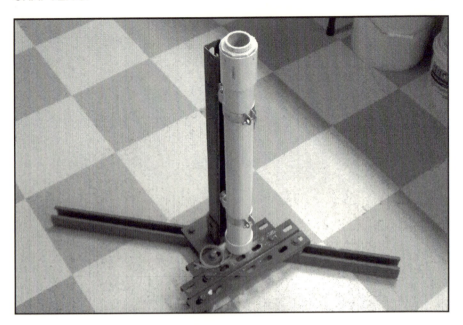

FLAMMABLE

Step 05
IN A REMOTE SAFE AREA – Put the K450 engine on the test stand with the nozzle facing up. The engine should slide up and down on the strut channel easily and without restriction.

THE ABOVE PICTURE ILLUSTRATES THIS STEP FOR REFERENCE PURPOSES ONLY. MAKE SURE THIS TEST IS DONE IS A REMOTE SAFE AREA.

Materials
  • None

Tools
  • None

FLAMMABLE

Step 06
IN A REMOTE SAFE AREA – Place the hydraulic scale under the end cap.

THE ABOVE PICTURE ILLUSTRATES THIS STEP FOR REFERENCE
PURPOSES ONLY. MAKE SURE THIS TEST IS DONE IS A REMOTE
SAFE AREA.

Materials
- None

Tools
- None

Step 07
IN A REMOTE SAFE AREA – Set up a video camera on a tripod at a safe distance so that the camera will record the gauge movements. Start the video camera recording.

THE ABOVE PICTURE ILLUSTRATES THIS STEP FOR REFERENCE PURPOSES ONLY. MAKE SURE THIS TEST IS DONE IS A REMOTE SAFE AREA.

Materials
- None

Tools
- Video camera
- Tripod

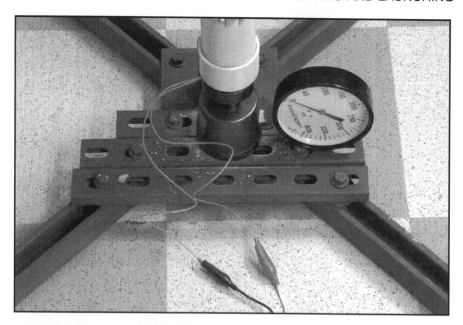

Step 08
Purchase a 100' long extension cord. Cut off both ends, strip the wires and install alligator clips on the black and white wires.

IN A REMOTE SAFE AREA –Test one end of the 100' extension cord with a multi-meter to ensure there is no power present. Hook up the alligator clips to the igniter leads.

THE ABOVE PICTURE ILLUSTRATES THIS STEP FOR REFERENCE PURPOSES ONLY. MAKE SURE THIS TEST IS DONE IS A REMOTE SAFE AREA.

Materials
- 100' Extension cord with alligator clips attached at both ends

Tools
- Multi-meter

Step 09
IN A REMOTE SAFE AREA – Use the 100' extension cord to provide power from a 12 volt automotive or marine battery to the ignition wires and the engine will fire.

THE ABOVE PICTURE ILLUSTRATES THIS STEP FOR REFERENCE PURPOSES ONLY. MAKE SURE THIS TEST IS DONE IS A REMOTE SAFE AREA.

Materials
- 100' Extension cord with alligator clips attached at both ends

Tools
- 12 Volt automotive or marine battery

Step 10
IN A REMOTE SAFE AREA – Stop the video camera after the engine has fired so the data can be analyzed later. Remove the engine casing from the test stand after it has had a chance to cool.

THE ABOVE PICTURE ILLUSTRATES THIS STEP FOR REFERENCE PURPOSES ONLY. MAKE SURE THIS TEST IS DONE IS A REMOTE SAFE AREA.

Materials
• None

Tools
• None

K450 DETAILS

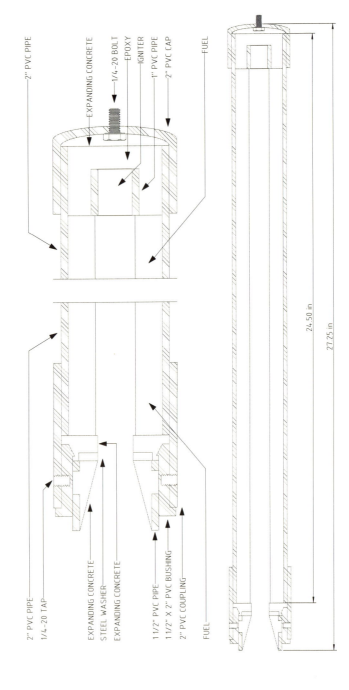

2" PVC PIPE

EXPANDING CONCRETE
1/4–20 BOLT
EPOXY
IGNITER
1" PVC PIPE
2" PVC CAP

FUEL

2" PVC PIPE
1/4–20 TAP

EXPANDING CONCRETE
STEEL WASHER
EXPANDING CONCRETE

1 1/2" PVC PIPE
1 1/2" X 2" PVC BUSHING
2" PVC COUPLING

FUEL

24.50 in

27.25 in

PARACHOOT

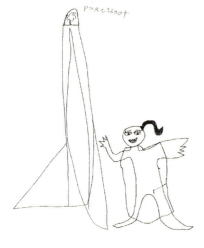

35254142R00109

Made in the USA
Lexington, KY
04 September 2014